BLUFF YOUR WAY IN HI-FI

SUE HUDSON
&
JOHN CRABBE

Ravette London

Published by Ravette Limited
3 Glenside Estate, Star Road,
Partridge Green, Horsham,
Sussex RH13 8RA
(0403) 710392

Series Editor – Anne Tauté

Cover Design – Jim Wire
Typesetting – Input Typesetting Ltd
Origination – Acorn Litho Colour Printers Ltd
Printing & Binding – Cox & Wyman Ltd
Production – Oval Projects Ltd

The Bluffer's Guides are based on
an original idea by Peter Wolfe.

CONTENTS

INTRODUCTION

What Hi-Fi is

Hi-fi stands for high fidelity. Some radicals argue that absolute fidelity is a bad thing, but that is super-bluffing – too sophisticated for beginners.

The fidelity, or faithfulness, which complex hi-fi machinery is designed to have, is to the *original* sound – that which you would have heard in the studio or concert hall at the time of recording. This is the cause of a lot of misunderstandings and frustrations, as hi-fi addicts up and down the country wear themselves, their soldering-irons and their bank-balances thread-bare trying to make a vicious little studio with dodgy mikes sound like the Albert Hall. This, of course, is NO-fi at all, which just goes to show what a very inaccurate name our subject bears.

It is considered *infra dig* to refer to your equipment as 'the stereo'. The correct term is 'my hi-fi', or even better – 'my system'. Never, ever, mention the Japanese – except with a sneer. The Japanese do not make hi-fi, but ear-wounding **rack** and **midi systems** and the happily fading music-centre. At the more bohemian parties at the Consumers' Association, grown men have been known to admit that, in fact, the Japs make some pretty good stuff nowadays – a match for many a European version. But this is naked anti-bluff, to be treated with the distain accorded to Galileo by all those well-fed Cardinals.

Since it is the flashy looks and conveniently small, value-for-money character of equipment that attracts the man in the street, real hi-fi goes out of its way to be very plain, large and expensive. In this way, you can show a perfectionist disregard for any considerations other than sound quality. And hi-fi *is* a search for perfection, so however good your system, you must never be satisfied with it (unless you are a manufacturer). The law of diminishing returns holds no terror for you. Whatever it costs and no matter how ugly, it's worth it. You dream at night of vast ribbon speakers, red-hot super-amps, and turntables made of gold and marble that cost more than your house.

Perfectionism inevitably leads to elitism and snobbery, and the new bluffer should pick his way very carefully through the hi-fi hierarchy. This starts from the top with the owner of the very latest esoteric state-of-the-art system, who looks down upon the modern classics super-fi user, who in his turn feels superior to the people who have hung on too long to the old classics, who seem enthusiast perfectionists to the mid-fi Volvo set. They find even the best Japanese racks and midis horribly vulgar, while those who buy them look down their noses at the cheaper Jap boxes and music-centres. But the music-centre is delux to the porters of 'ghetto blasters'. In turn, these are hi-tech to the flat-earthers who still play records on their old Dansette and tune into the Light Programme; and in the Third World there are probably places where a Dansette is a status symbol.

STEREO

Stereo is an illusion. This is not mystical observation but psycho-acoustic fact. For mystics of course, psycho-acoustic facts are also an illusion. Real life seldom produces identical sounds in two places at once. If it does, and you happen to be standing exactly between them, the sound will seem to be dead-centre. The middle Beverly sister must have had this illusion all the time. This is the way stereo works, and with variations on the theme of 'right-hand down a bit', it can move the sounds around between the speakers. Make some sounds a little quieter and more reverberant than the others and you have some semblance of depth too. All very clever.

When first introduced in the late '50s, there were the inevitable controversies, with some quite respectable people claiming that stereo actually worsened the sound. Try turning your system to mono (if it has a switch) and see if you agree. Fortunately most saw that stereo was a major improvement, and some wild-haired enthusiasts got carried away and said it made further development pointless. For instant *déjà vu* see Dastardly Digits.

The first stereo demonstration discs were Monty-Python affairs that had steam trains thundering through front parlours up and down the land, to amazed audiences. The ping-pong match was the classic cut – hence the expression 'ping-pong stereo'. Super-bluffers know who won. Ultra-bluffers can tell the make of the bats and ball.

A small proportion of the population cannot hear stereo. This makes a good bluff, since claiming you are one of a small band is always an advantage (unless the small band has a rare skin disease). Not being duped by an illusion is also impressive, and it's impossible for anyone to call your bluff. As most people, even hi-fi experts, do not bother with the finer points of stereo – since it demands that you put your chair (called a stereo seat) in unlikely places which are never near the fire or the coffee table – becoming a stereo connoisseur is also good bluffing. Always aim for the extremes.

Phase

Phase is a problem associated with stereo, and the subject of heated debate. If the music seems to be climbing up the wall, lacks bass, and you can't locate any individual sounds, your speakers are probably **out of phase**. Reverse the connecting wires in the back of one of your speakers to correct this odd fault, caused by invisible battles in your room, where sound waves are cancelling each other instead of merging to produce stereo. They do this because two wrongs *do* make a right – or rather, one big healthy wrong – and a wrong plus a right of equal moral strength makes nothing at all in the end (like stealing a fiver and then giving it back again). Connecting one speaker the opposite way round to the other induces speaker psychopathology, where one's wrong is the other's right and vice versa.

The abitrary nature of speaker morality is demonstrated by the fact that if you wire up the other speaker the same way, everything works all right again. There are some people, though, who claim that hi-fi morality is absolute and that speaker pairs wired up back-to-front are merely twisted and evil in unison and that this is audible. This eternal ethical problem seems no nearer an agreed solution in hi-fi than it is in philosophy.

In the meantime, arguments over **absolute phase** as this disputed effect is called – or if you need to be ultra-correct **absolute polarity** – are fine opportunities for showing-off, one-upping, bluffing, and generally behaving rather badly.

POWER

As in politics and business, this is ultimately what matters most in hi-fi. It's measured in **watts**, of which you ought to have several hundred per channel if you wish to cut much ice. Otherwise you can cheat by doubling whatever you've got (for stereo), then doubling it again (for peak as opposed to mean power), then increasing it by about 50 per cent (for short-term music power). This will convert a 50 watt amplifier into a 300 watt model. Never forget there are lies, damned lies, and power ratings.

EQUIPMENT

If you are to bluff successfully in hi-fi you must have *some* idea of what is going on technically, although it is gratifying just how little real knowledge you actually need. Many a hi-fi pontiff cannot read circuit diagrams and doesn't really understand the basic electronics.

Happily, more important than eye-glazing details of inner-workings is your knowledge of the relative *caché* of types and accessories. It is this aspect you are advised to concentrate upon – sprinkling a few hard facts in thin slivers over the subject, like white truffle over rissoto.

Let us start with what is inelegantly referred to as the 'front end'.

Turntables

Turntables are relatively simple devices whose job it is to spin the record at the right speed and not jiggle about or hum in the process. This they do simply (and badly) causing **wow and flutter** when the speed varies, rumbling, and making the needle jump when you sneeze. It is essential that your turntable is dedicated to $33^1/_3$rpm, with 45rpm only available after dismantling the deck and fumbling with belts and cogs. This is your introduction to the **Inconvenience Rule**, in which all the extra useful facilities that are on offer are frowned upon and said to erode sound quality. This rule means that such restful devices as auto-cues, automatic arms or – horror of horrors – record stackers, should never be seen on your deck, or advocated, except in play. The irony is that the 45rpm speed is capable of much better sounds than $33^1/_3$, as devotees of 12-inch singles will know. In these hi-tech days it will come as a shock to young students of hi-fi to learn that the central working part of most decks is a rubber band.

An essential bluff turntable accessory is the **record clamp**, which is fitted over the middle of the record via the spindle. Whether these make any worthwhile improvement to the sound is, like so much in hi-fi, a moot point. But it *is* certain that the act of doggedly screwing and unscrewing the thing at every change of sides is very professional.

On this theme, we recommend a complex disc-play ritual involving the full gamut of special cleaning brushes for record and stylus, **anti-static guns**, and a selection of mats. To avoid ruining the effect, you should know the right and the wrong way to handle your records. The right way starts with washing your hands – or ostentatiously donning a pair of surgical gloves. You then slip one hand inside the record sleeve, resting the label on your middle finger and the edge on your thumb, tip and slide the record out. The wrong way involves grasping the main body of the disc with sweaty, jammy fingers, dropping cigarette ash on it, and leaving it lying about – preferably on top of more nude discs. This method is excellent as a shock tactic for perfectionist visitors.

A refinement of this tactic is to have a meticulously clean disc to hand which is the one you *really* play. The resulting pin-drop silent background should soon have others thoroughly demoralised.

Arms

The **pickup arm** is a terrible nuisance. It doesn't actually participate in the action except to add unwanted vibrations, although in far-off brassy days it was designed to carry *wanted* vibrations from the soundbox behind the needle to the horn, hence the alternative name – tone-arm – still anachronistically used today. Its sole purpose nowadays is to hold the cartridge over the record and take its wires to the back of the deck.

Early arms used to be massive affairs that had to be wrestled with. Then they went to the other extreme and the breeze from a passing butterfly could move them. The pendulum swung back to high-mass arms for a while and has now settled somewhere in the middle. Weedy, low-mass arms match well with very **compliant** cartridges which are easily led, while more stiff-necked cartridges need the strong-arm tactics of a heavier mate to get them working well. A 'mis-match' (good word) can cause terrible embarrassment.

Considering its passive role, the arm can and should look suprisingly complicated, with balancing weights and threads hanging off the end in what laymen call 'a mess'. The linear tracking arms so popular in cheap and nasty systems are rare in real hi-fi. They appear only at the very top end of the market where the necessary precision engineering is available to make this basically more correct system work. Occasionally, a revolutionary arm will make an appearance on the scene. These should be regarded with grave suspicion.

Cartridges

The **cartridge**, with its stylus, is a crucial link in the hi-fi chain. People may, and of course do, argue about the importance of tone-arm excellence, but all would agree that there are significant differences between cartridges. It is important to remember that the continuum of excellence is exactly matched by a continuum of expense. So wide-spread is this coincidence, that when it is broken, manufacturers altruistically double the price of the odd good-though-cheap item so that the public doesn't get confused. A key point to be clued-up on is the **moving-coil** versus **moving-magnet** debate. These alternative types vie for precedence, with moving-coils being top-dog at the moment. Ceramic cartridges (see Dansette) are always good for a laugh.

One characteristic of cartridges seldom talked of (so

bluffers should talk of it often) is their temperature dependence. In short, they don't like the cold, mistracking and getting chilblains on their **response curve** when exposed to the average British front parlor in February. If you live in a draughty mansion, instal a light bulb over your cartridge's resting place to keep it warm. Friends will be terribly impressed.

The real heart of the cartridge is its stylus, or **needle**. (To call it a needle shows you are on relaxed good terms with the subject, like calling a violinist a fiddler). This must be up to scratch. Which means it must be a 'naked' diamond, very, very small, and **elliptical**.* Indeed, some pernickety hobbyists have a microscope for viewing the little beasts, to check for wear.

Equal bluff-points go to the people whose sensitivities are fine enough to detect wear in its early stages with the naked ear.

You should also know how to listen for stylus/cartridge problems by playing the appropriate tracks of a record. If, on loud bits, it sounds like Tina Turner gargling glass, you have **tracking** distortion. If, on high pitched inside tracks it sounds like Tina Turner eating cotton wool and broken glass casserole, it's **tracing** distortion. Casual hi-fi owners confuse these two, so appear confident that you know which is which, even if you don't. For extra panache, drop into the conversation the fact that they are respectively mechanical and geometrical in origin – although you will need some ingenuity to do this with the right throw-away air.

* The tip of the system in low-fi is usually spherical, but in hi-fi elliptical types (or more correctly bi-radial) are the norm. They can have exotic names like *Shibata*, *van den Hul* or *Pramanik*, but the most unusual (not to say unheard of) is the *Apocalyptical* stylus – said to be marvellous for playing the last act of *Gotterdammerung*.

The furious wiggling of the stylus by the record grooves (in two directions at once for stereo) moves the magnet or coil in the cartridge, which is a dolls-house-size generator and so translates the squiggles into electrical signals and thence into obscurity. Which brings us to. . .

Amplifiers

If you take the electrical signal straight from the cartridge to the speakers, you'll hear nothing at all. There is not enough power in it to move the cones. So the job of **amplifiers** is to put lots of power (watts) into the signal – or rather the other way round. Imagine that the electrical signal is Adolf Hitler when he was just an embittered house painter, and the power in the mains supply is Germany between the wars. Sensing that something is in the air, lots of Germans stop rushing around in opposite directions (AC) and queue up in neat rows in Nuremberg (DC in the power supply) to hear him speak. Hitler then fires the crowd with new purpose, so that instead of just marching into Warsaw on his own, the whole crowd goes out and annexes Poland. That's how amplifiers work.

As well as being electrically obscure, amplifiers are as class ridden as their owners. The aristocracy is naturally Class-A. These are expensive, overweight, self-indulgent objects that use up a lot of energy – especially the valve versions, which glow with satisfaction. As in life, most amps are Class-B (working-class) or A/B (middle-class). These have to concentrate hard to speak with upper-class accents, keeping slimmer and cooler in the proces. Class-Ds are bohemian oddities.

As amplifiers exist to add power to the signal, the

more watts the merrier. The record to date is 300 watts per channel. The bluffer's king of power amps, the **Krell** (200 watts), is named after a race of extinct heavyweight aliens, the core of whose planet was a gigantic power station, and who listened to strange electronic music. Amps named B&Q mk 5811 seem very pale by comparison.

Pre-amps

Pre-amps or control units, conform to the minimalist maxim 'less is more'. Indeed, with the advent of Compact Disc, it is possible and desirable to by-pass them altogether. The pre-amp corrects the incoming signals from all those 'front ends' which arrive in various states of unsuitability for straightforward power-amplification. They also give you a chance to mess around with the sound, although in practice this is limited by current hi-fi philosophy. Apart from inputs for phono (American for gram), tuner and one auxiliary, purist pre-amps may carry only a volume control. Muting, balance controls for adjustable stereo, and tone-controls are all damned under the Inconvenience Rule.

Japanese pre-amps are a study in what to avoid in the way of gadgets, which tells us that graphic equalisers are right out – although a £9,000 tone-control has recently appeared which is deemed acceptable, perhaps heralding a change in fashion. Quad owners have always rejoiced unashamedly in their well-considered filters. But then Quad owners are a mite too sensible for bluffing emulation. If you *must* have controls, they should be **passive**, not **active**. These terms describe complex internal processes best left to the imagination.

Head-amps

Head-amps do the same thing (boost puny moving-coil output from deck) as pre-amps, only more conspicuously.

Tuners

Tuners are both too simple (functionally) and too obscure (technically) for amateur bluffing. The emphasis in modern hi-fi thinking is very much on records (and, grudgingly, CD) to the almost complete exclusion of radio and tape. The main point to settle with tuners is the nomenclature. Tuner is the simple main-stream term, but you may choose from: FM tuner, AM/FM tuner, radio tuner, plain radio, or even wireless (once short for 'telegraphy without wires'). Also, if you dare to flout the Inconvenience Rule by combining the tuner with an amplifier, there are receivers or tuner-amps.

That makes eight names altogether, not counting 'tranny', 'ghetto-blaster' and 'Brixton briefcase'.

The hi-fi significance of radio lies in the excellence of the classical music broadcasts by the BBC. These are received on VHF (short for Very High Fi) otherwise known as FM (Fine Music) and will provide you with as many string quartets and endless minor song-cycles as you can take. If you happen to want rock during the day, cricket, or Radio 4 on school-days, you'll have to listen on AM (Awful Mess) and wince and bear it. Test Match Special sounds best on Bakelite valve sets which claim to receive Hilversum and the Home Service anyway.

Should you feel undressed without some technical

terms, 'superhet', 'repeat-spot-suppression' and 'phase-locked-loop' are good for casually dropping into the conversation.

If radio is a neglected aspect of hi-fi, then aerials are even more recherché. But do not overlook the bluff possiblities of the obscure. In the unlikely event that a hi-fi friend plays you his FM tuner, ask if he has noticed a touch of 'multi-path', and suggest that a multi-element aerial array would do wonders for his 'signal-to-noise-ratio', rather than the homely 'H' you noticed on the chimney on your way in.

Compact Disc Players

Compact disc players are either the best thing ever or a travesty of the real thing, depending which team you are on. For more see Dastardly Digits. After the nit-picking complexity of record-playing gear you might innocently assume that hi-fi nuts would welcome the simplicity of the new medium with sighs of relief. But that is not the way the enthusiast mind works at all, and many long-suffering veterans of the LP feel distinctly restless around CD and have already started breaking into the machines and tweaking around, replacing bits and rewiring.

Technically, CD is deeply satisfying and clever. Neither is it as difficult to understand as you might expect. Despite this, most hi-fi experts haven't a clue how it works and come out with the most glorious fundamental bloomers. Take the trouble to learn the rudiments, and you'll be able to call the bluff of many a conventionally based buff.

Basically, it works just like a computer, with the on-off information (pits in the silver disc) read by a laser.

As the laser light is reflected back in-focus (no pit) or out-of-focus (pit) the digital signal is transmitted via a lens, which 'sees' the light beam. The laser travels across the disc so fast that Wagner's entire *Ring Cycle* would last a mere 1½ hours if played on LPs at the same speed (which some would say was a good thing too).

Playing starts in the middle and works towards the edge – it is essential to remember this when looking for scratches and pin-holes in the right places. Pin-holes cause the most problems. They are introduced during pressing, and are neither the fault of the record shop nor induced by careless handling at wild parties. They are made visible by holding the disc up to the light – which gives you a good flourish to use at your next wild party.

The advertisements for CDs misled the public into thinking that neither dirt, scratches, ring-marks, nor even direct nuclear attack, could affect their sound quality. This simply is not true, and the bluffer should therefore over-compensate by making just as much fuss over cleaning and handling his silver discs as he does over his black ones. It is essential to know how to open the notoriously tenacious CD jewel cases, and to master the trick of wresting the disc from the central spigot without bending it into a Frisbee – experiment in the privacy of your own home.

Courtesy of a few super-chips, the digital code is instantly translated back to normal – analogue – electrical signals, and before you can say 'glitch' these are zapped with filters, anti-aliasing devices, de–emphasisers and mini-amps, etc. The *really* clever stuff happens when the message understandably arrives a bit garbled. Our cunning machine can actually work out what the real message is – like reading bad morse-code, e.g. D-w-o-n- w-i-t-h H-i-l-t-e-r. 'Interleaving' is a good word to use here. The thoroughness of this error correction is the main difference between players, and most of the CD 'tweaks' so far, involve taking the strain off these circuits by reducing vibrations.

The more 'bits' of information one has about the signal the better, so 16 bits are inherently better than 14 bits. But you can get round this by looking at the bits you do have, very, very carefully. This is called over-sampling. First generation players either had one or the other. Once enthusiasts had splashed out on these, the manufacturers brought out 'new, improved' models which have both. Which just goes to show that CD marketing men are every bit as clever as CD inventors.

Cassette Players

Cassette players are only of prime interest to engineers, for whom the medium's wide range of problems are a reliable source of amusement. For the rest of us they are more an object of convenience than of hi-fi sanctity. Buy the best you can by all means, but it must be admitted that the sound quality of the cassette medium, even after decades of virtuoso tweaking by designers, is low-grade in comparison with the other 'front-ends'. Thousand pound decks *do* go a long way, but the owner of these will gain a reputation as an eccentric boffin rather than a main-stream hi-fier.

Of all the troubles besetting the average deck, **hiss** is king. Some years ago, a certain **Ray Dolby** from America came over to England, studied at Cambridge and then gave to a grateful world the Dolby noise-reduction system. This ingenious invention reduced the loud hissing on tape machines to tolerable levels by a selective **pre** and **de-emphasis** (jokes about taking the hiss are banned). It's awfully clever, and you may pass a pleasant few hours one rainy Sunday trying to understand it. That was Dolby 'B', now we have 'C' as well, which is more thorough.

High frequencies have a rough time of it too, what with being 'squashed' and disappearing when the deck is 'out of Azimuth' – good phrase this, you can't go wrong using it when you think a cassette lacks a certain brightness. Then there are wow and flutter (generically known as wobble), ALCs, AVCs, AGCs (automatic level, volume, or gain controls), HX to deal with HF squashing, and bias switches – for, not content with the complexity of the basic machine, inventors have come up with three different types of tape to play on them. You should always try for three 'heads', and of course two heads are better than one. Finally, some machines pump and breath the hiss, which sounds like a snake on elastic.

So there is a lot to learn, or ignore, depending on which way you look at it. Fortunately, it looks as though in the near future we won't have to bother. The advent of portable and in-car CD players, R-Dat, and hi-fi video recorders (which perform better than even the most expensive cassette deck, record *Dynasty*, have a longer playing time *and* cost half the price of anything approaching the same quality), sounds the knell of doom for old ways.

Those wanting a crash-course in cassette nasties before it's too late, should buy themselves the cheapest *Walkman* they can find in Woolworths. And a packet of aspirin.

Cables

For decades, **cables** were just bits of wire to string equipment together. Apart from tripping up the odd enthusiast and thus smashing piles of treasured LPs, they weren't thought of as important at all. Then came the tweakers. Now cables are seen as crucial components, making all the difference to a system costing thousands.

To confuse the issue further, there are two camps among cable advocates:

1. Those who are all for single-strand wires
2. Those who advocate very thick cable with hundreds of strands

It doesn't matter which you choose, but remember always – the copper with which the wires are made must be **single crystal**. This is an expensive way of producing wire, said to be audibly better. Rich hi-fi owners are born with silver cable in their sockets – oxygen-free of course. If you just don't care, use 7-amp mains cable. If you *do*, then it's best to get some advice from an unbiased cable supplier (for names see *Utopia News*) and go for dedicated cables for each interconnect, paying as much as you can't afford.

What you *do* with your cable once you have got it home is also crucial. One band of tweakers hang their cables on bits of string suspended from the ceiling. If this seems a trifle eccentric, it does make sense to take field-effects from under-floor mains supplies into account when laying your wires.

Perhaps the most outré cable controversy revolves around the idea that they sound different one way round rather than the other. The practice of finding out which way is 'up' is called **sexing**, and should be undertaken only by consenting adults. Scientific backup of dubious relevance has been submitted. The theory says that, because of crystal boundary effects and skin effects, etc. the path of the electrons is easier in one direction. It is essential to get this all sorted out so that the electrons know whether they are coming or going.

Cable effects have now entered into the hi-fi volcabulary, and one description of a good audio system is 'short'. This is because with so many distortions per inch, the best piece of wire is a short one, thus minimising the degradations of a component for so long thought to be beyond reproach.

Loudspeakers

By the same basic process as in cartridges only on a bigger scale, **loudspeakers** translate the electrical signal (by now very powerful indeed, and ready to take over the world) back into mechanical movements which move the air with cones, ribbons or diaphrams. This may sound crude, but it's just the way, give or take the odd wind section, that sounds are made in nature. The problems with speakers all stem from the fact that we are trying to reproduce instruments by playing them back through another instrument (the speaker) which has its own sound. This is known as **coloration**, which because the Americans first used the word, is spelt in the American way, without the 'u'.

Some methods of moving air have less coloration than others. A general rule-of-thumb is: the less thick, heavy and generally bulky the 'transducer' the less the coloration. Very small boxes with thick drive units sound quite nasty, although not as nasty as they once did, as great strides forward have been made here in recent years. This you must readily admit in hi-fi circles, while giving the impression that you wouldn't have one in the house.

The perfect speaker would have no mass and no dimensions. The perfect speaker does not exist, and if it did, it still wouldn't.

The nearest to perfection yet devised are the 'panel type' speakers. These come in various forms, including the electrostatic, the ribbon and the magneplanar. These make music enthusiastically described as being 'transparent', 'neutral' and 'uncannily natural'. To a man they take up far too much room and get claustrophobic within three feet of a wall. The Inconvenience Rule is all-powerful.

Ears

Fiendishly clever devices for turning air waves back into electrical signals, which the brain 'sees' as beautiful music. Nobody really understands how ears work, so anyone claiming to do so can safely be ignored. Many psychological points can be scored by the young bluffer (and the female bluffer of any age) who stresses his/her natural high frequency acuity. You find the line-time-base on TVs quite intolerable (15.625 kiloHertz) and the pilot-tone on stereo broadcasts deafening (19 kiloHertz) *when not adequately filtered* – be careful here. The higher the **Hertz** the better. Remember no-one can ever hear much above 20 kiloHertz, so you must feel very confident to claim that you can.

People over forty (especially men) inevitably lose their hearing at these stratospheric sonic heights, but you may still impress by knowing *exactly* where your sensitivity falls off, e.g. 'I roll off at 12 ½ right ear and 14 ¼ the left nowadays'. These declarations need not be accurate, as very few people have access to the specialist audiometers necessary for meaningful readings.

Another factor of hearing sensitivity is that of relative sound levels, expressed in **decibels (dB)**.

Decibels are very confusing, partly because they involve horribly complicated things like logarithms and also because they are not supposed to be an absolute measure of anything, but an indication of how much louder or quieter something is than something else. Thus two flutes are 3dB louder than one, but so are two symphony orchestras.

It's best simply to remember that:

a) any normal person can detect a change of 1dB
b) someone who claims to hear a change of 0.1dB is probably a liar
c) anyone who can't hear 3dB has cloth ears.

For convenience, an absolute scale of decibels is commonly agreed upon that runs from 0 to 125 dB. Those wishing to experience 0dB must gain access to a sound-proof room. After half an hour, when your body noises have subsided, you will be able to hear the sound of air molecules striking your ear-drums. At 125dB you start to feel pain.

In the concert hall, music of very wide dynamic range extends from about 30dB (a good deal quieter than your sitting room) to 100dB (which would wake up grandma). So you can see that high concert-hall fidelity is socially dangerous, a fact to be firmly ignored by bluffers, who should insist they need their 100dB of hi-fi flesh.

PUTTING IT ALL TOGETHER

However much you know your way around the jungle, there is still the subtle art of equipment usage to be explored.

The first point to be clear on is that any hi-fi system worth bragging about will be an uncoordinated assortment of the best in each field, with all the visual appeal of odd socks. The turntable may be Scottish, the amplifier American, and the speakers English. Although market pressures are changing the picture a little now, it's still true that a custom-made, neatly stacked system is the sign of the hi-fi philistine.

Next on the agenda is the **warm-up**. This phrase, in a surprising breach of jargon etiquette, means just what it says. Plug in and switch on at least an hour before attempting to play music. When in doubt allow even longer. Class-A amplifiers, true to their profligate natures, are best left on all the time.

You may run into **matching** problems where the output from your CD player into the pre-amp is too strong and the signal distorts and clips. Conversely, some cartridges have outputs that are too weak and feeble (see Head-amps), and panel speakers present some very odd 'loads' to amplifiers. These loads are expressed in ohms, which, because it is all Greek to most people, are written thus Ω.

All the plugs and sockets round the back should be easily get-attable, so arrange your equipment accordingly. This is designed to be awkward – the Inconvenience Rule again – but worth it for bluff-opportunities. Overawe lesser mortals by bragging that you remove and reinsert every plug at least once a week to keep the contacts clean (have a tin of Duraglit nearby) – but *not* if you have gold-plated connections.

In which case you'll still want to have them easily accessible, this time for showing-off.

Positioning is of vital importance, and in order to show that you are aware of this you should furnish your music-room sparsely, with the chairs and sofas obviously slung in wherever there was room *after* you have found just the right spot for your speakers. Only titchy speakers should have their backs to the wall, in defiance of criticism and to improve – or rather boost – the bass. More confident noise makers should take their rightful place in the middle of the room. In any room there is one optimum place where your speakers sound best. This fact is extremely inconvenient, but that should come as no surprise by now. Wherever you keep it, each hi-fi item has a stand of hi-tech design custom-built for it. These stands are useful visible testaments to your know-how – buy lots. Resonant hollow objects like redundant speakers, sideboards and empty-headed friends theoretically play havoc with sound-waves, so ban them or stuff them with old hi-fi magazines.

If you are going tweaking, this will merely be the starting point of an elaborate and conspicuous feather bedding of your system (see Tweak Freaks). But one tweak that most have taken up is removing the grilles from the front of the loudspeakers. This makes them look uglier and more stridently technical, which is good. It's also meant to improve the high frequencies.

Fixing It

Such complex and expensive equipment is bound to be temperamental, and when it *does* go wrong, you should turn first to the two great maxims of engineering troubleshooting:

1. Try turning it upside-down
2. Thump it

The 'inversion' method works well for moving-coil loudspeakers, but is not recommended for turntables or CD players. The 'velocity-impact' method is more subtle than you might think: hi-fi experts know exactly where and how hard to hit wayward machinery. There is even a set of IEC regulation hammers for the purpose.

Trying to remove hum is guaranteed to drive you to the edge of insanity. Even experts can spend whole days on this.

Buying It

There are hundreds of shops all selling the same items and not selling the same items – just like Oxford Street shoe shops. The assistants are as thick as Monster cable; you might as well stay at home.

Real hi-fi is sold in rare specialist shops which advertise in the magazines. These are Aladdin's caves full of esoteric marvels, and all have the essential **demonstration room**. This is a blissful place that you may book for an hour or two and in which you finally get to listen to all the stuff you've heard so much about.

The catch is that you are expected to buy it afterwards. Assistants in these shops are often unbearably condescending. To out-bluff such people you need to look round and see what top-class equipment is *not* there. Wait for an assistant to come to you, and ask for it thus: "Do you have the *Apogee Scintillas?* I hear great things of them." (This with much enthusiasm.)

'No, but we *do stock the Mageneplanar IIIs*, they are superb.' At this, look as crestfallen as you can and reply "Oh, no." It is now quite safe to ask for an audition of the equipment you are *really* interested in – always with a hint of disappointment in the voice.

Some success has been had with a short black leather skirt and fish-net tights. Those lacking the nerve or the legs for such a gambit could always carry this book as an unspoken declaration that you know all the moves.

HOW TO LISTEN

There was a suggestion that ear-plugs should be worn for twenty-four hours before listening. This was meant to improve sensitivity, but although it would work, it shows the wrong attitude. In fact, most '**golden ears**' take no noticeable care of their hearing – exposing their ears regularly to head-banging levels of rock, and screaming editors. Truly effective listening is an altogether more showy affair, involving much physical effort (a warm-up can be beneficial for humans as well as for hi-fi).

The basic position is, of course, exactly equidistant from and in front of the speakers. This is merely a starting stance, however, and experts are soon to be observed bobbing and swaying, standing on tip-toe and grovelling on the floor, always with an alert expression. Soon they grow more restless and go roaming around the room listening for 'eigentones' and clapping their hands to test for 'flutter echo'. More relaxed exponents may leave the room altogether to facilitate peripheral hearing. Very relaxed audiophiles often leave the whole house – finding the snug room in the local pub strangely revealing.

Make a reputation for yourself as a serious and acute 'listener' known as a golden ears, by either:

a) refusing all alcohol before listening (if those about you are knocking it back) on the grounds that it clouds the judgement, or
b) getting legless (if the others are abstemious) on the grounds that relaxation is essential for fine perception.

Provided you steer clear of blind tests, which show up inconsistencies of judgement, the sky's the limit.

When questioned, sensible bluffers should not babble about bass extension or treble clarity like a text-book. Instead, you must aim for the poetically vague. Cogitate gravely and come out with phrases like 'translucent pools of velvety ambience' or 'coherent inter-transient dynamics'. These have the right sense of meaningless precision. For more examples, read the less reputable magazines. 'Dimensionality' is a good and fashionable buzz word, describing the degree of realism, definition and spaciousness of the stereo picture. Becoming a stereo connoisseur is good bluffing.

Finally, do stress the nature of the recordings used for listening. Good equipment *reveals*, so you should point out subtle and usually ignored sonic details such as the contents of the conductor's pockets, or the age of the guitar strings. Say you found 'that crude edit' outrageous – a safe ploy since most recordings are riddled with edits and it would take a brave person to contradict you.

The Test

Should you be involved in an equipment listening test in a friend's house or hi-fi shop dem room, you will need further guidance. Everyone will be practising basic listening skills, so you need a few more refinements.

1. You must be the first to spot any obvious mistakes such as the speakers being out of phase, or left-right reversed. Anyone who is anyone will spot these fairly soon, so an instantaneous response is vital.
2. Next, make a fuss about some aspect of the room –

the light bulbs are ringing, or the stereo is vague. If you are dogmatic enough, nobody will be inclined to disagree and you will have scored several million bluff points.

3. Reject at least one record on the grounds that it is not of high enough quality. Old-scholars should reject any recording which is not totally digital, while young-pretenders should condemn all digital manifestations as being a 'cacophonous, tuneless drone'.
4. Spend ages getting the volume level *exactly* right.

Once actually listening, the well timed gurgle of irrepressible amusement can be devastating in its power of suggestion. And, if you can get away with it, the guffaw is irresistible. Otherwise, keep very quiet and look absorbed. Afterwards, you might slip into the conversation such remarks as: 'fine ambient resolution', 'horribly plummy coloration', 'a papery top', and the all-conquering, 'disturbingly neutral'. The less-than-100 per cent confident should take their lead from the others, joining in by being more damning or enthusiastic than the others as soon as possible. No one ever said bluffing was a gentlemanly pursuit.

HI-FI TYPES

It is essential to decide which branch of hi-fi you are best suited for, and stick to it. Inconsistency of approach means instant loss of credibility. This is because the game is played by teams, each of which hates the other teams and accuses them of selling-out, of cloth-ears, or of having IQs of $33^1/_3$. Below, we catalogue the main bluffs for your selection.

DIYers

DIYers are a dying race. Once a thriving group wooed by all the journals and manufacturers, they are now forced to buy esoteric American magazines and to cannibalise old equipment for parts. The modern availability of relatively inexpensive and adequate hi-fi has left the field to the true enthusiast, ever in search of the ultimate audiophile system on the cheap. A difficult bluff unless you have vast electronic and engineering knowledge and lots of spare time. Hardly anyone builds their own loudspeakers nowadays (see Nostalgia), cassette players are next to impossible, and turntables and arms, let alone cartridges, always were a craftsmanlike pursuit. That leaves tuners, which are so well made by the professionals that it's not worth the bother.

Amplifiers, now looking a lot less pretty than in their brassy acoustic days, should be huge and glow in the dark, or have capacious heat-sinks to dissipate the colossal power throbbing through their circuits. If you try to buy one of these brutes it'll cost you the price of a new family car. Make one and it'll cost you the price of a rusty old Mini. The catch is, that unless you really know what you are doing, it'll *sound* like a rusty old Mini.

Habits and Habitat: Their rooms are congenitally untidy but quite cosy. Nothing matches. There are hundreds of trailing wires, solder blobs cemented to the carpet, burn marks from the soldering iron on the table, bits of plastic (semi-conductors) and wire-ends under foot. The bought equipment is identical to the Old School (see below) with the addition of a *Connoisseur* turntable – once available in kits. The rest has the lo-tech, hand crafted look of the home-made: nothing is ever quite finished; plastic knobs are much in evidence. It'll probably sound wonderful, except for the nasty hum they can't seem to get rid of and the fact that you have to keep banging the pre-amp to get the left speaker to come alive. Key names are *Linsley-Hood*, *Nelson Pass* and *Nelson Jones* (designers), *GEC* (valves), *Radio Spares* and *Maplin* (parts).

DIYers play organ music very loud, frequency-band records, and the climaxes of hugh choral works, over and over.

Tweak-Freaks

These are impoverished enthusiasts without the education for full-scale DIYing. The tweaker lives in a rosy world where, by a series of bizarre and relatively cheap ritual peripheral changes, a nasty little midi-system can be made to sound like a million dollars. A strange mixture of fact and fiction, some tweaks are just sound engineering practice that most people (especially engineers) can't be bothered with. Others fly in the face of known physics.

This is huge fun, and the pages of the hi-fi journals are presently full of black-belt bluffers making spectacular reputations using the 'Einstein was wrong' approach.

The current description for the improvement brought about by all this fuss and bother is 'a veil was removed between myself and the music'. Tweaking can therefore be facetiously described as the dance of the seven thousand veils.

Habits and Habitat: The tweaker's room will have an odd hi-tech untidyness. Do *not* offer to help with the housework and move all the cables out of the way or pick up the steel lumps so carelessly left lying around on top of the equipment. For the one thing you could *not* accuse a tweaker of is carelessness (unless it's with the laws of physics): everything is there for a reason. Smile approvingly at the holes cut in the carpet (to allow speakers to 'couple' with the floor), grunt knowingly at the nasty gunge oozing out of the audio plugs. That old bowl of water on the radiator is there for humidity control, not for the cat.

To show you are really *au fait* with the trends, a good ploy is to offer to remove your digital watch and leave it outside. This will gain you a quick reputation as a serious and meticulous seeker after perfection. Restrain yourself from laughing when asked if you have tried sexing your cables, or you will spoil everything. Just keep telling yourself 'they all laughed at Christopher Columbus'.

Key names are *Monster* and *van den Hul* (cables), *Sound Organisation* (stands) and **Martin Colloms** and **Jimmy Hughes** (tweak scribes). They play jazz, MOR rock (especially Dire Straits) and the odd flute Concerto.

The Old-Scholar

The old-scholar is the direct opposite of a tweak freak. Where tweakers take the approach known as 'subjectivity', i.e. if it *sounds* good it works, the old school were brought up on the 'objective' viewpoint, i.e. if it *measures* good it works. To them, tweaks are either old-hat or insanity. A constant source of annoyance to old-scholars is the general reluctance of subjectivists to prove that the effects they hear are more than self-delusion by taking part in **blind** tests, the sneaky practice where, with the equipment hidden behind blinds, no change is made at all and listeners swear blind that several veils have been removed. Even more annoying for subjectivists is the refusal by objectivists to accept the results of the few blind tests that have been done which show that there *is* something in some tweaks after all.

To carry off this pose successfully, you must have either great credibility as a scientist/engineer, or grey hair and a Quad amplifier. Arch old-scholars have all three. They also have the most enjoyable social lives. You can spot an old-scholar easily at a party – he is the one telling jokes and drinking rather a lot. Move closer and you will hear music cropping up in the conversation.

The old-scholar is sure of his ground, successful, prosperous, and believes in enjoying himself. In contrast, the tweakers look twitchy, are abstemious and *never* crack jokes. Hardened cases appear to have no sense of humour at all. For them life is earnest, bitter, and the grave is the goal. They know that it is impossible to be sure of your ground on anything, and even more impossible to make a living out of hi-fi nowadays.

Habits and Habitat: Old-scholar's hi-fi is deliberately unfashionable. It is very well set up within reason, and sounds good, but never spectacular. The music room will be cosy and conventionally furnished – there is a lot of wood around. Key names are *KEF, Spendor, Quad, B&W, Thorens, Shure, Tandberg, Revox* and *SME*. They play Mozart, Beethoven and the romantics.

B&O People

B&O people are an odd sub-species, not really in the mainstream of hi-fi, but still managing to make others feel uncomfortable. 'Should appeal to B&O people' is a phrase reviewers use every time this style-conscious Danish equipment turns up. But we have never seen any attempted definition of this term. So who are these people and why are they different?

The really awful characteristic of the type is its implacable cool: the implication that everything is, of course, of the best and that it is vulgar to make any fuss about it. They take good sound quality for granted and bother themselves with other, equally important matters such as looks and ergonomic excellence. You may truthfully criticise Bang & Olufsen sound as not always being first-class, but the B&Oers will remain unflustered. They know that there are far more important things a hi-fi system can say about its owner than how discriminating his ears are. It is better to go with the flow, and instead talk of music or design philosophy. Drop a few design names – like *Yamahas'* ahead-of-its-time cassette deck, or *Gales'* beautiful and discontinued skeletal turntable.

Habits and Habitat: The house is impecably decorated in the modern style. A Le Corbusier chair stands in the corner. It is severely tidy. They keep a cat. B&Oers are naturally drawn to the clean, fuss-free modernism of CD, of which they already have an impressive collection. The music on these discs is avant-garde or Classical with a capital C. They find the Romantics too noisy and self-indulgent. Pop is much in evidence, with cool off-beam stylists like Talking Heads and Sade to the fore.

Bass-Freaks

Bass-freaks are a jolly band, committed to the search for the perfect reproduced bottom-C organ pedal note of 16.35 Hertz. Once an ordinary all-round enthusiast, the bass freak has never recovered from a visit to the Albert Hall when the seats shook to the power of the great organ. Life then became a love affair with huge horns and flapping woofers. Houses are rebuilt to accommodate massive hidden DIY horns and they suffer agonies of indecision with crossovers and eigentones.

Rock and roll has given us a sub-species, equally obsessed, this time with the 'tunefulness' of bass guitars.

It is easy to be seduced by the sonic pleasures of good deep bass, since it seems to have a magical effect on the poorest hi-fi, warming and giving body to the thinnest of sounds. It is both exhilarating and relaxing at the same time. Indeed, we are aware of incipient bass-freakery in ourselves.

You could try deflating the bass-freak by observing that most music carries bass only down to around 60 Hertz – but it seems a shame.

Habits and Habitat: The house is large and detached. Furnishing will be big and cosy and mid-way between old and new. Rampant slovenliness is not unknown. Expansiveness and generosity are the hallmarks of the bass freak's hospitality. Key names are *Baldock* and *Crabbe* for DIY horns, *Tannoy* and *Klipschorn* for off-the-peg horns, and *Linn Isobariks* for the new wave. The music you will hear will be either Widor, Bach or Tchaikovsky's *1812* for organ music and cannon fire, or Power Station, Led Zeplin and Van Halen for drilling bass guitar.

Stereo Obsessives

Although stereo is an illusion, it can sometimes be uncannily precise, provided you are prepared to sit with your head rigidly held at exactly equal distances from each loudspeaker, or wear headphones which have the performers on a line passing through your brain. But there's a species of hi-fi fanatic whose dream is to free us from the tyranny of the 'stereo seat' so that we can all experience stable, pin-point imagery no matter how many people are listening. That such people are often bachelors and usually listen to music alone is odd.

Whatever psychoanalysts might make of them, such stereo purists can be very intimidating, with their talk of speaker radiation zones, head diffraction effects, localising cues, and so on. If you should ever be caught in the web of such a perfectionist, it's a safe bet that he's going to use you as a guinea-pig.

There are two opposing bluffs for this. One is to pretend that you are a member of that 10 per cent or so of people who can't localise stereo images properly.

The other is much more effective. You start by citing the teachings of the late **Joseph Enock** (a pioneer stereo obsessive and early tweaker), who maintained that any worthwhile system should work equally well if the listener faces *away* from the speakers. Accordingly, insist on turning your back on the hi-fi before passing judgement. Follow this up by throwing yourself flat on your back and lying alternately with your feet towards and away from the speakers. Explain that you are acting on some seminal BBC research which revealed that reflections from the floor aid stereo perception. It might even be true.

Nothing arouses the passions of stereo obsessives more than mention of **Dr. Amar Bose**, so use his name with discretion. This American manufacturer makes **omnidirectional** loudspeakers which spray the sound out in many directions. Many hi-fi people think it sounds wrong and gives rotten stereo. But most Americans think it's great and Dr Bose's much-maligned boxes are top sellers there. Another key name is **Blumlein** – pioneer of stereo recording and inventor of the purist miking technique known as the 'crossed pair'.

Their music is likely to be wind quintets by unknown East European composers.

Rich and Vulgar

These people may have everything you always lusted after, but probably gained it too easily. The merest suggestion that they do not really appreciate the quality of their system will keep up your morale. It's likely that something has gone radically wrong with it somewhere due to the careless, offhand attitude of the rich mentality – giving you scope for impressing by transforming the sound with a few basic adjustments. The music will be of such triumphantly MOR artists as Barry Manilow and James Last.

Just Rich

Those who are just rich not only have the best, but *do* appreciate it. They tend it with loving care and precision engineering. This sort probably dine regularly with Von Karajan and have a genuine love for exquisite musical byways. There is no way the rest of us can keep up – let alone compete. Just remember that they are so insufferably perfect that everybody else does just what you do: keep away.

Cheap and Cheerful

This is the low-brow approach to hi-fi. The cheap and cheerful brigade do not pretend that their music-centres and personal stereos are the last word, but are so relaxed about it that they make tweaky perfectionists feel distinctly heavy-handed. This they do by playing good music all the time and thoroughly enjoying it.

If you are attracted to this point of view yourself (an understandable reaction), remember that it is essential to project the right sort of nonchalance. While it is fine to enthuse over the latest gadgetry from Japan, you do so because it is clever or useful – you are not in awe of it. You acknowledge that there *is* wonderful hi-fi out there, but cheerfully admit that you just can't be bothered with it. Mention casually that you have a season ticket for the Proms and are going to Bayreuth next year, and you are home free.

CONTROVERSIES

The one thing audiophiles *would* agree is that anyone who says anything about hi-fi will instantly find himself surrounded by arm-waving protagonists. Argument and dissent inhabit the subject the way fleas inhabit a cat.

Even in the gentlemanly days of wind-up gramophones, angry fanatics in mutton-chop whiskers and top-hats got all fired up over questions of horn shapes and needle types. Later on, in the '60s, when transistors replaced the old glow-in-the-dark valves, a new kind of nasty called 'crossover distortion' occurred which no one thought was audible. But many people *could* hear it, and said transistors sounded terrible. Innocents would be surprised just how long it took the designers to stop saying rude things about these 'subjectivists' and get on with some investigating. Once they found a measurement that showed masses of distortion at low levels – just those quiet moments in music where you hear nasties best – the designers were happy and fixed the problem.

This basic scenario has been played out ever since, whenever a new idea is introduced. Bluffers must therefore decide on their stance for once and future controversies. The objectivist (measurement-fixated designers) must be able to exude strong silent authority with ease and be happy to remain above it all. This is restful but dull, with few bluff opportunities.

We recommend that you go for the subjectivist (ear-fixated experimenters) approach in a big way. This gives you the freedom to range at will over all aspects of hi-fi, pontificating, eulogising, damning and using bluff words to great effect. Find your own particular hobby-horse and go on about it just a touch too much.

The Great Amp Debate

An unresolved conflict which started in the early '70s. It finds subjectivists insisting that quite apart from crossover distortion, some amplifiers sound better than others, and objectivists saying that they can't because they all measure the same. Going on past history, and considering that the design and components of amplifiers can differ radically, the subjectivists probably have a point.

If a subjectivist, you should have firm opinions about the amps you prefer and use vivid language to describe their superiority. Spend most of your time listening to amps and debating with friends over your next 'upgrading'. Valve amps are much loved by subjectivists and have made a great come-back under their wing in recent years. Objectivists will tell you that valves sound good (if they do) because of 'soft clipping'. Ignore them.

Dastardly Digits

The advent of Compact Disc opened the floodgates to subjectivist arrogance. To a man they hated its alien hi-tech claimed perfection, the only variety being in the vehemence of their hate and the eloquence of their invective. Some claimed to be able to hear the steps in the digital encoding, others said it bore the same relation to music as hamburger did to beefsteak – which was a nice analogy even if it *was* a misguided judgement. The old-scholars all loved it, declaring that the system was indeed perfect and that any sonic raspberries arose because CDs revealed 'nasties' on the master-tapes which LPs glossed over. Having said this, they then retired from the scene, refusing to argue about it, which aggravated the subjectivists so much they got really abusive. A lot of two-digit signals got sent between the camps.

Much to the disgust of subjectivists, the general public seemed to have a taste for hamburger and bought the new technology in encouraging numbers. A not-so-subtle retrenching took place where the second generation players were judged to be improved and almost acceptable. Since then 'tweaked' British versions of the Philip's machines and overpriced Japanese models have gained tacit approval. Some of CD's harshest erstwhile critics are now earning a nice living reviewing players and discs.

Because of this it is probably wise nowadays to stay quiet when asked about the relative merits of CD and LP, but you are on safe ground praising CD's silent background and convenience. The differences between CD players are very subtle and should therefore be exaggerated by those posing as golden ears.

Tweakery

This is subjectivism in full flight and an inspiring sight for bluffers. It's really a new science – aural engineering – which takes hi-fi back to basics with its concern over the behaviour of wires and capacitors, etc, but instead of measuring these building-blocks electronically, actually listens to them.

The trouble with this method is that the ears are directly connected to the brain, so that even though testers' ears may be fine, their minds can get them into all sorts of scrapes. Listening is naturally a highly selective process, so inattention, bias and preconceived impressions – along with the great difficulty of finding perfect recordings with which the tester is familiar – make the listening test problematic. Minute technical slips can also create false effects.

The situation horrifies traditional scientists, who know from long experience how easy it is to produce invalid test results. They keep asking for double-blind **A/B tests** which exert some degree of control over this wayward state of affairs. And tweakers keep not doing them, on the understandable grounds that such tests are always run by objectivists who are openly hostile to the concept of tweakery. There is much mistrust on both sides, and you should be careful not to treat the whole thing too flippantly or you will be surprised to find you have stirred up unexpected depths of emotion.

Manufactuers simply *love* tweaks. They generate demand for ready-made versions of gadgets and modified equipment thought up by tweak scribes. It is now possible to buy 'bricks' of just the right weight and composition to put on top of your hi-fi. Esoteric cable made of copper possessing alchemical properties can be yours for a few pounds per foot, and gold is liberally spread over everything, from cable terminals to the humble mains plug.

Although some tweaks are feasible, if radical, others seem nearer to witchcraft and are hard to swallow. If you accept the tweak philosophy it is impossible to discriminate between them all, which is alarming, and also makes the cynic very distrustful when confronted by yet another tweak accessory 'A snip at £50!' What with capacitor distortion, foils for CDs, stands, mats, special paste for annointing connections (actually *called* Tweek) and many, many more, it's reaching saturation point. Even dedicated enthusiasts are finding it hard to keep up. But don't be daunted, it's the basic approach that counts, and if someone starts discussing a tweak you haven't heard of you may show your independence of spirit by poo-pooing it.

One word of warning to objectivists – be *very* careful about tweaks. Science is nowhere near the exact and correct process they tell you it is in school, and at least one loony aberration will be accepted fact in future years.

The Right Stuff

Brand loyalty is very strong in hi-fi, and whatever you choose will be despised by someone. There are two ways round this:

1. Pretend that you are in-between systems and making do while deciding what to buy next. This starts an enjoyable debate, with advisers topping each other in ambition without it costing you a penny. This process is known as **up-grading**.

2. Go for the most respected main-stream system and defend it on grounds of sanity. Such a combination would have a *Lin Sondex* turntable with their arm and one of their cartridges, and *KEF* or *Quad* speakers. These makes have stood the test of time and survive in the critical environment of hi-fi because the men behind the names are brilliant engineers who still care about quality and innovation.

At the cheaper end, Quad amps are still held in esteem by most, but you will be expected to spend disproportionate amounts on amps by tweaky friends. *Nakamichi* is *the* name in cassettes. CD players are tricky: it will be ages before the market settles down and one model really stands out. *Philips* are the best bet, since they are not Japanese and invented CD in the first place and so have the technological resources to make fundamental improvements. Their first player, the *CD100*, already looks like a classic.

That lot will cost you in the region of £4,500 with a tuner thrown in. But this is a mere blip on the scope compared with the most expensive system in the world – at £105,000.

Nostalgia

Just as some people collect old masters instead of modern art, you too may play safe and collect classic hi-fi rather than go to all the trouble of sorting out the good from the bad in modern design. Surround your *arriviste* self with the aura of Old knowledge in the form of *Radford* valve amps – still sought after and fetching handsome prices second-hand. *Revox* tape recorders are all-time classics, as are *Thorens* record decks and *Tandberg* cassette players. *Quad* is a reliable name and all their old products have cult followings. Valves have recently made a comeback in modern amplifier design and are really in vogue at the moment. Horns, the great '60s obsession, are way overdue for revival. Start building now.

MUSIC

The holy grail of hi-fi. Since critics of the subject used to be forever remarking how little actual music we played on our perfect systems, hi-fiers lay great stress on their love of the art nowadays.

This is very handy for the newcomer, since it provides a ready cover-up for awkward moments when the conversation gets beyond him. Simply divert attention from the equipment to the music it is playing, thus: 'You know, this really is a terrific/awful performance – reminds me of the great occasion at the. . .' and so on.

There are records and *records* – the right ones being **DMM (direct metal mastered)** and **direct-cuts** (see Glossary) available at head-spinning prices at hi-fi shops. Pressings made in Japan or Germany fetch twice the normal price because of their supposed superiority, operating on the well-known maxim 'British is worst'.

It is now essential for the serious enthusiast to have a large LP/CD collection of respected music. Masses of test records and brass-band albums simply will not do. Second-hand shops and jumble sales are handy for acquiring acres of vinyl, only a small percentage of which need actually be playable.

Out of his capacious library, the hi-fi freak only plays a few well-loved discs. Prominent among these will be the very same records that he used to play before the cultural guilt set in: thunderous organ works for bass, choral works for HF intermodulation, and test records. A single favourite track will turn up again and again for 'checking' innovations and new equipment.

HISTORY

It all started with *Mary had a Little Lamb*, which is the first recording ever made. Scratched by a needle into tin-foil and played back in exactly the same way, it sounds like W. C. Fields in a sand storm, instead of **Thomas Alva Edison** in his laboratory. From this it can be deduced that early hi-fi was nothing of the sort, which is perfectly all right since nobody claimed it was.

Such was the novelty of the invention that some innocent enthusiasts declared that it was just like real life – much as the audiences at the first cinema shows thought the actors were behind the screen. The sounds made by dragging a thick needle across wiggly lines were 'amplified' by horns made of brass. This made early machinery very good to look at and worth a fortune in antique shops.

Typically, the French insist that their inventor, a Monsieur **Charles Cros**, created the gramophone first (by three months). And with Gallic egocentricity they still celebrate his breakthrough with an annual gramophone award.

This period – roughly from 1880 to 1930 – is called the acoustic period. The music was captured, stored and replayed by direct mechanical means. Even the turntable had to be hand-cranked, giving music-lovers some healthy exercise and a sense of involvement. After that, we have the electronic age, which can be split into the valve, transistor, and digital eras.

Although hardly anyone seriously claims that acoustic machines are superior, it is the done thing, and profitable bluffing, to proclaim that valve technology has never been bettered (or equalled) and that transistors, and especially digits, are instruments of the devil.

Failures and Oddities

Heath-Robinson is the patron saint of hi-fi, and his more devout followers have paid homage to him by inventing some truly silly devices, most (but not all) of which have gone to the great dem-room in the sky.

Front-End Follies

The problems involved in rescuing the music from record grooves have inspired more bizarre contraptions than any other province of hi-fi. Let us start with *Transcriptor's Transcriber* turntable.

This had a fixed arm which, instead of moving over the surface of the LP, stayed resolutely still while the entire bulky platter trundled towards it on runners – a case of the mountain coming to Mohammed.

Even more extraordinary was *Audio-Technica's* weightlessness-ready turntable which, instead of relying on good old gravity to hold the stylus in the groove, fought it all the way by pushing the platter up onto the arm. This design was a real collector's piece since it included two other over-ambitious features: a linear tracking arm and a vacuum platter.

There was also a brief moment in hi-fi history when everyone wanted his system to **suck**. LPs *will* wobble annoyingly while being played, and although the modern record clamp is a partial solution to this problem, a much more thorough method is to suck it onto the platter. Unfortunately the vacuum was produced by turning relatively elegant turntables into chunky metallic warhorses. Worse still, the pumps were noisy.

The *Groovac* was a mini vacuum-cleaner which hung over one's LP, sucking up all the grime in the grooves, so instantly reducing 'background noise' – all the hiss and crackle on discs. Sadly, no-one knows if it worked, because it made such a din you couldn't hear the silence.

Some ideas were noble in execution but addled in conception – such as the *Garrard Music Recovery Module.* A complex electronic device again designed to remove background noise, it was inclined to remove chunks of music and replace them with silence. People noticed.

Ever in search of a successor to the arm, lateral-thinking inventors – at *Sony* this time – road-tested their *Sound Wagon* in the early eighties. This was a model VW van with a stylus on its undercarriage which drove round the LP at *roughly* the right speed. But at least it was portable, which brings us to the Soundburger.

Ever since the demise of the wind-up gramophone, the great picnicking public has been in sore need of a portable LP playing system. Thanks to *Audio Technica's Soundburger* they could once again stuff their favourite crooners into the hamper and take to the hills.

Ignoring minor necessities like support for the disc, and leaving the LP naked under attack from pork-pie crumbs, sand and wasps, it was actually praised by design buffs (so much for beauty through utility).

Back-End Bloomers

Loudspeakers had their finest hour with *Yamaha's* legendary ear-shaped drive-unit. About as sensible as a fork shaped like a mouth and as alluring as a scent-bottle shaped like a nose, someone at Yamaha clearly hadn't quite grasped the rudiments of acoustics. It did, however, have a good stereo image: as enthusiasts exclaimed, 'The orchestra is spread from ear to ear!'

Armchair experts raved over the *Acoustic chair* from *S J Court*. A fireside comfy chair with stereo in the wings and a bass port in the seat, at last heavy metal *could* grab you by the seat of your pants.

The Acoustic chair was only one of an honourable series of wonderful ways to disguise ugly speakers. Others include, speakers in a drinks cabinet, in a giant egg, in a kitchen table, even in spitoons. But foremost in the affections of the hi-fier are the *Jordan-Watts* pots. These look like early Eutruscan olive-oil jars and are a must for old hippies.

All Rounders

In the early eighties **Hugo Zuccarelli** made quite a media splash with his *Holophonics*. Amid a flurry of secret patents and weird theories he seemed set to become a millionaire.

Poor fellow, he should have known that surround-sound systems have tried and failed to get off the ground as often as a lead turkey. Manifesting itself under multitudinous guises such as *Hafler*, *Quadraphonics, Ambisonics, SQ, QS* and *CD-4*, the basic idea was the same – to get the music coming at you from all sides rather than just in front. Regardless of what you might have thought, this is more natural and so higher fi. The main problem was that they all involved the expense and domestic upheaval of installing at least two extra speakers (except Holophonics). The ambient information was stored on the LP in code (except Holophonics) which, due to the suicidal incompatability tendencies of hi-fi, came in four different flavours, all needing different black-boxes to decode them (except Holophonics). The public rejected the whole project in exasperation (including Holophonics).

Nevertheless, it makes the music so much more

enjoyable, expansive, concrete, etc., that it is very difficult not to get carried away and start installing it yourself. Restraint is essential, however, since it is terribly *de trop* in hi-fi circles and the mere mention of it will elicit groans all round, thus giving a superb demonstration of real-life surround-sound. Brave bluffers hitch their surround-sound star to hopes of digital revival.

Last Word

Finally, honourable mention must go to *Hi-Fi News* for their Flux-Shunter. The ultimate tweak, this was a scale model of the great pyramid of Cheops whose pyramid power was reported to keep styli sharp. It caused cancelled subscriptions and outraged letters from old-scholars who hadn't noticed that it was advertised in the April issue.

MAGAZINES

There are only two classic journals in the field: *Hi-Fi News* and *The Gramophone*. The rest are upstarts that come and go with the whims of fashion, keep changing their names, and are always incorporating each other.

In recent times *Hi-Fi News* has grown a bit feverish with tweak disease, but now shows signs of cooling. *The Gramophone* is so old and set in its ways that, like the royal family, it exists above controversy. Its continual harping about music leaves little room for technical debate – definitely *the* rag for old scholars. Tweak freaks get *Hi-Fi Answers* for bigger and better questions and *HFN* for technical tweak details from Martin Colloms and its Accessory Club – source of all that is fine and good, and of annoying Christmas presents for old-scholar friends. Old-scholars read *HFN* for Angus McKenzie, a much-loved and respected radio and tape expert. Both this and *Hi-Fi Answers* are worth reading purely for such masters of the pen as Ken Kessler. This is symptomatic of their general drift from engineering to journalism.

New Hi-Fi Sound has a growing readership, seduced by its sense and thoroughness. Once just plain *Hi-Fi Sound*, perhaps in the fullness of time it will be renamed yet again as *Old Hi-Fi Sound*.

All four still operate on the assumption, once well founded, that British hi-fi is the best in the world. They rarely review Japanese equipment, unless it is a CD or cassette player, preferring to concentrate on what they call 'the cutting edge of technology', all of which seems to be extremely expensive. It also all seems to come from America, which is the *real* home of hi-fi these days.

This is why DIYers buy American magazines,

namely *Audio Amateur* and *Speaker Builder*. These are penny-plain affairs full of real engineering and things to build – just like *Wireless World* and *HFN* in the good old days. English designers are still thinking up good ideas, but like all good ideas today, they go to America to prosper. This gives rise to the odd situation of having to ship a journal all the way from the USA to find out what old Reg up the road is doing.

A new journal, with the plebeian name *What Hi-Fi?*, has burst upon us which actually talks about the sort of hi-fi that most people buy, in plain English. It already has an enviably high circulation, which just goes to show that all the average music lover wants is a quiet life and a healthy bank balance.

But the biggest selling hi-fi reviewing magazine in the country is *Which?* In amongst the washing machines and steam irons, are technically sound, thoroughly blind-tested reports on low- and mid-fi and Compact Disc, all reduced to a few blobs and a couple of words. These are so sane that everyone in hi-fi ignores them.

When Hi-Fi? and *Where Hi-Fi?* collapsed from lack of interest. *Why Hi-Fi?* never got off the ground.

SOME USEFUL PHRASES

To tide you over gaps in the conversation when you are obviously expected to come out with something impressive, here are some ready-made and fool-proof phrases.

But that isn't psycho-acoustically valid.

Your eigentones are bunched too close together.

I'm afraid your phase and amplitude components are incompatible.

I'm sure there's a temporal displacement on the eigen-vector of the principal horn.

Those down-band spuriae add a pleasant crunchiness to the upper-mid.

On *my* speakers, you can hear a tube train pass under the Kingsway Hall at that point.

When did you last have your stylus profile examined?

The inter-sample dither is rather over-quantized, but what else can one expect from these thin cables?

My system is *so* revealing I can only bear to listen to three records these days.

(On scrutinising the back) Ah, as I thought, the *old* model.

Have you tried putting a 100 kHz square-wave through this?

And the all-time classic:

You've done very well to make this system sound so good.

GLOSSARY

Hi-fi talk is built upon obscure jargon. At the last count there were well over a thousand special terms in 'common use'. But most of these are used only by genuine experts and can be ignored for bluffing purposes. Drop a few of the following carefully selected 47 key words in the right places, and everyone will assume that you also know the meaning of the other 953.

Absolute phase – The belief that a blow is a blow and a suck is a suck, and that some people can tell if a particular recording should be blown and sucked or sucked and blown.

Ambience – A useful general-purpose word which enables you to remark that it is absent, excessive, unnatural, dry, wet, bright, dull, cavernous, deep, shallow, or (rarely) just right.

Arrival-time – The precise moment when you hear the beginning of a British Rail announcement that a train is running late. Same at both ears if facing the loudspeaker.

Bias compensation – Technique whereby a pickup's tendency to swing to the left (when viewed from the front) is counteracted by a force to the right. If you find this politically offensive, simply view your pickup from the rear (or purchase a model made by the Alliance Manufacturing Co., which is claimed to be free of left/right bias).

Bits – Small chunks of digital music (6dB wide) found queueing up in CD players awaiting processing. There should be 16 bits for faultless reproduction, like Mr Gladstone's 32 chews for faultless digestion

– hence the expression 'getting the bit between one's teeth'.

Breathing – Audible effect generated by poor noise-reduction systems. Much easier to detect if you stop doing it.

Bulk eraser – Device used by keen recordists to prevent their tapes from growing fat in middle-age.

Clipping distortion – When an amplifier has had just about as much as it can take. It responds by giving the listener a clip round the ear.

Compression – What makes music of impossibly wide dynamic range suitable for listening to at home.

Crossed pair – Stereo microphone arrangement beloved of purists. It's always acceptable to remark that a good recording sounds as if it were made using this method – even though it almost certainly wasn't.

Crossover distortion – That which causes sound-waves to hang around doing nothing when they should be moving on to positive or negative regions. Found in bad Class-B amplifiers.

Crossover frequency - The number of times one needed to cross the room to change 78rpm record sides when playing a complete opera. Now used in connection with loudpeakers.

Crosstalk – A double-act by stereo pairs, where one partner repeats everything the other says in a quiet voice. This is such a bad act that the performers subsequently have terrible trouble with their image.

Damping – Method of putting out fires in hi-fi equipment. The trouble is that damping is inclined to affect wanted as well as unwanted fires. It's a complex subject, full understanding of which requires one to go on a Damp Course.

Direct-cut – Recording in which performers play directly as the disc is cut. This replaces the inaudible effect of the tape-recording stage by the audible effects of a poor performance.

Dither – State of mind of subjectivist journalist on being confronted by a 'perfect medium'. Applied to CD players to improve their sound.

Doppler effect – Familiar to laymen as the change of pitch when a police-car siren passes by at 80mph on a motorway. In hi-fi, a similar effect is heard on music as loudspeakers repeatedly rush towards and away from the listener.

Double-blind – Type of listening test in which no-one knows what's happening or what anyone else is doing. Said to be necessary for obtaining scientific results.

Eigentones – The ghosts of dead bass-freaks which lurk in the corners of music rooms and moan in response to favourite frequencies. When over-excited they also stand up and wave (see Standing-waves).

Gain – Profit margin of hi-fi shop.

Glitch – Effect heard when a Compact Disc player is hit by rampaging turntable manufacturer. Sounds rather like a glitch.

Hertz(Hz) – Small pains which occur once per second. Thousands of them are so painful that they almost kill you – hence kilo-Hertz(kHz).

Hole-in-the-middle – Stereo defect most easily detected by listeners with cold white ears in mint condition.

Hysteresis – State of near panic in a tape recorder when magnetic particles on the tape don't know whether they're coming or going.

Infinite Baffle (IB) – Type of loudspeaker which functions in a way that no-one has yet been able to fathom, probably because they are sealed boxes.

Insertion loss – A sexual term which has crept into audio via the Freudian symbolism of plugs and sockets (see also Rise-time).

Intermodulation – Incestuous goings-on between related musical sounds while passing through amplifiers. Results in offspring called IM products.

Labyrinth – Type of loudspeaker enclosure made by Daedalus Manufacturing Co., powered by an inbuilt minotaur.

Loudness control – Bossy volume control which also changes bass and treble whether you like it or not, because that's how it thinks your ears work even though they don't.

Moment of inertia – That instant in time when one can't be bothered to get up and turn over a record. Proportional to one's weight multiplied by the square of the distance between self and record player.

Muting – Facility for cutting the tongue out of a radio tuner.

Objectivism – School of hi-fi philosophy which objects

to everything about subjectivism. Objectivists don't believe their ears if they hear anything they can't measure, and don't make any measurements that might support what subjectivists say.

Presence – Characteristic of reproduced music which makes it seem very near. Many recordings seem so near that they could do with a dose of Absence.

Radiogram – Hi-fi's equivalent of apspidestra, chamber-pot and antimacassar.

Rise-time – Time taken for one audio component to respond to another. Depends on the degree compliance.

Slew rate – Speed with which an amplifier swings sideways in order to avoid very fast on-coming signals.

Standing-wave – Situation where reflected sound-waves arrive back in the same place at the same time. This makes for overcrowding, hence the expression 'standing room only'.

Subjectivism – The belief that the way to tell what a piece of equipment sounds like is to listen to it. This is considered far-fetched whimsy by objectivists.

Super-tweeter – Very expensive breed of canary found in the most ambitious loudspeakers. Used for reproducing bat-calls, pilot-tones, and other HF sounds which most people wouldn't want to hear even if they could.

Surround-sound – Any scheme for making sound come at you from all directions. Ideal for reproducing the atmosphere of noisy swimming baths.

Thermal runaway – Situation where the output transistors of a power amp decide to do a bunk when things get too hot.

Tweeter – Smallest drive-unit in all but the most expensive loudspeakers, used for reproducing bird song. Most sound more like parrots than nightingales.

Upgrading – Process by which hi-fi enthusiasts build lusty over-drafts. Involves replacing equipment claimed to be 'the best arm in the world' or 'the closest approach to the original sound' with something better.

Upper-middle – A term from Marxist sociology which has drifted into hi fi. Hence the common complaint that a loudspeaker has too much or too little of it, depending on whether the model is above or below one's price range.

Valve amplifiers – Outlet for hot air.

Woofer – Largest drive-unit in a loudspeaker, produces growling bass. Inferior types only manage to yap, but better models will Bach at organ music.

Wow and flutter – Unaccountably funny sounds like 'boing', 'ttthhwwwpp' and 'pop' which turn grave and beautiful piano concertos and woodwind recitals into music-hall acts. Most at home on tape machines and the sound tracks of American soap operas, where their querulous ups and downs are highly appropriate.

THE AUTHORS

Sue Hudson showed early signs of engineering genius when she made her first printed-circuit at the age of eight, and some good swopsies in the playground with jewellery fashioned from wire and resistors. Having missed the electronics lessons at school when forced to do needlework, she now makes up for it by writing about music, reviewing for a Hi-Fi magazine and sitting on the *Which?* expert listening panel (which can be very lumpy). Entirely surrounded by men in her professional capacity, she derives great pleasure from her work.

She is married to a brilliant electronic engineer, four oscilloscopes, two concrete horns, and eighteen experimental power amps.

John Crabbe started off as a radar technician, then bluffed his way into phonetics research by pretending to understand acoustics. He tried his hand at writing about washing machines, and finally annexed the editorial chair of a well-known Hi-Fi magazine which he occupied for 18 years. Pausing briefly to bluff his way into farming, he claims to be the only audio editor ever to take his own pigs to market.

He has gone through various manic DIY phases, leaving behind a succession of houses with vast horn loudspeakers built into the walls, and found time between moves to invent a device now used on millions of pickup arms (which did not earn him a penny) and to write *Hi-Fi in the Home* (which did).

THE BLUFFER'S GUIDES

Available now @ £1.00 each:

Accountancy	Music
Antiques	Paris
Class	Philosophy
Computers	Sex
Consultancy	Teaching
Golf	Television
Hi-Fi	Theatre
Hollywood	Wine
Management	

Coming, September & October 1987:

Feminism	Bluffing
Jazz	Marketing
Literature	Photography
Modern Art	Publishing

All these books are available at your local bookshop or newsagent, or can be ordered direct from the publisher. Just tick the titles you require and fill in the form below. Prices and availability subject to change without notice.

Ravette Limited, 3 Glenside Estate, Star Road, Partridge Green, Horsham, West Sussex RH13 8RA

Please send a cheque or postal order, and allow the following for postage and packing. UK 25p for one book and 10p for each additional book ordered.

Name...

Address...

....

THE BLUFFER'S GUIDES

In preparation:

Advertising
Architecture
Astrology
Ballet
Bank Managers
Beliefs
The Body
Cinema
The Classics
Defence
Espionage
Finance
Gambling
High Society
Journalism
Law
Millionaires
Opera
Politics
Property
Psychiatry

Public Relations
Secret Societies
Selling
Ski-ing
Stocks & Shares
Travel
University
World Affairs

The Americans
The Australians
The British
The French
The Germans
The Japanese

Amsterdam
Berlin
Hong Kong
Moscow
New York